Praise for

50 Things You Can Do with Google Classroom

By Alice Keeler and Libbi Miller, Ed.D.

"*50 Things You Can Do With Google Classroom* by Alice Keeler and Libbi Miller is the definitive guide to bringing the social learning platform to life in your educational setting. Not only do you get a breakdown of all the great functionality of the tool, but they offer sound educational advice on why, with deep ties to pedagogy, to make the technology even more powerful."

—**ADAM BELLOW,** *Founder of eduTecher/eduClipper*

"In *50 Things You Can Do With Google Classroom*, Keeler and Miller have produced a powerful resource for a powerful tool. The straightforward directions for use, alongside concisely delivered philosophical frames for why one would try each idea, make for a strong overall picture of this tool. Many teachers need the comfort that comes from having a guide for newer tools. With so many schools adopting Google Apps for Education, this book on Classroom comes at the right time. "

—**RUSHTON HURLEY,** *Executive Director, Next Vista for Learning*

"Alice Keeler and Dr. Libbi Miller combine clear, easy-to-follow explanations with practical application ideas for using Google Classroom with students. This resource is a treasure for any teacher using Google Classroom!"

—**CATLIN TUCKER,** *Teacher, Trainer, Speaker, and Author*

"This is the definitive resource on Google Classroom! Alice and Libbi have packed this book with practical, relevant ideas for using Google Classroom in your classroom starting tomorrow!"

—**KYLE PACE,** *Instructional Technology Specialist/GCT*

"Keeler's and Miller's tips for Google Classroom can help the most reluctant ed-tech user feel secure. Comprehensive and easy to follow, the text is a must have manual."

—**STARR SACKSTEIN,** *Author and Education Innovator*

50

Google
Classroom

Alice Keeler **and Libbi Miller, Ed. D**

Published by Dave Burgess Consulting, Inc.
San Diego, CA

http://daveburgessconsulting.com

Cover Design by Genesis Kohler
Interior Design by My Writers' Connection

Library of Congress Control Number: 2015938340
Paperback ISBN: 978-0-9861554-2-0
Ebook ISBN: 978-0-9861554-3-7

First Printing: May 2015
Second Printing: March 2016

Contents

Foreword by Jonathan Rochelle ix

Introduction xi

Introduction to Google Classroom 1

Overview of Google Drive and Docs. 2

Getting Started 4

Teacher View: A Quick Tour 6

Student View: A Quick Tour 11

Google Classroom App 15

50 Things You Can Do With Google Classroom

1. Make Class Announcements. 17

2. Share Resources 18

3. Keep Multiple Files in an Assignment 19

4. Create a Lesson 19

5. Go Paperless 20

6. Easily View Student Submission 21

7. Simplify the Turn In Process 21

8. Protect Privacy. 22

9. Encourage Classroom Collaboration 22

10. Reduce Cheating 23

11. Create a Discussion 24

12. Organize Assignments with Due Dates. 25

13. Feedback Before Students Submit 26

14. Email Students. 27

15. Notify Students Who May Need Help 27

16. Assignment Q&A 28

17. Create an Ad Hoc Playlist. 28

18. Email Feedback 29

19. Create Folders 30

20. Link Directly to Student Work 31

21. Collect Data 31

22. Share with Multiple Classes 32

23. Collaborative Note-Taking 33

24. Display Student Work 33

25. One Student, One Slide 34

26. Target Parent Phone Calls 34

27. Polling 35

28. Share a Document with the Class 35

29. Know Who Edits a Collaborative Document 36

30. Link to a Website 37

31. Peer Feedback 37

32. After-Hours Help 38

33. Distribute Notes 39

34. Sharing Informal Learning 39

35. Email the Teacher 40

36. Eliminate Schlepping Papers Home 40

37. Student Projects 41

38. Have One Place for All Files 42

39. Document Digital Work 43

40. Students Create Google Docs 44

41. Clearly Identify Student Work 44

42. View Assignments 45

43. Collaborate with Peers (PLCs) 46

44. Virtual Office Hours 47

45. Virtual Faculty Meetings. 47

46. Streamline Counseling 48

47. Observe Another Classroom. 48

48. Watch Students Do Homework. 49

49. Share Student Samples 49

50. Provide Choices. 49

51. Reuse a Post 51

52. Use Google Forms 52

53. Create an Exit Ticket 52

54. View Student Work 54

55. View Submission History 54

Conclusion 57

Acknowledgments 59

Additional Resources 61

About the Authors 66

Foreword

Any successful product requires focus on the user: an understanding of who will use the product and how they will use it. When we built Google Classroom, this meant understanding teachers and students and how technology could be applied to their problems and opportunities. We were lucky to have some former teachers working on the product design and engineering within Google, and, more importantly, we teamed up with active teachers to understand their pains and their hopes for teaching more effectively and efficiently. We focused on how technology could save them time, so they could focus on their teaching and their students. We tested early versions of the product (long before it was ready) in real classrooms and got feedback from teachers and students on how to improve the product. We iterated over ten months, expanding the early test groups and improving the product with each cycle. We launched Google Classroom in August of 2014, and, while it has been tremendously useful for so many teachers and students, that was still just the start of our feedback and improvement cycle. Google Classroom continues to improve as we listen to feedback and find the most impactful changes we can make to help teachers and students.

But the most impactful positive force for teachers in applying Google Classroom and other ed-tech products has been their peers. The practicing teachers, who have become experts in applying technology and have become mentors for others, provide something valuable well beyond the tools... that is, practical advice, solutions to problems, and confidence. These teachers – these mentors – find solutions to real teaching problems and innovate new ways to do things using technology. More importantly, they share these solutions openly and always look for ways to improve. These teachers have become the backbone for the educator support networks on Twitter, Google Plus, and other online and in-person networks. They are what make Personal Learning Networks (PLNs) worthwhile for so many teachers.

Alice Keeler and Libbi Miller are two such expert mentors. They've applied technology to classroom instruction for a combined, thirty-plus years at the high school level and both now educate teachers in the use of technology at the university level. That's their day job. But even when they're not teaching, they're still teaching. Alice's blog, TeacherTech, is well known as a source of solutions across many teaching topics and many technologies. Alice often gains her inspiration from questions that come from her Professional Learning Network (PLN). Since August of 2014, I have watched her blog become a go-to resource for expert advice about Google Classroom. For no matter how much we try to improve the product, there is no substitute for the advice of a practitioner. It's not only about simply learning the mechanics of a software product but also about the practical application of that product to a real need. Alice and Libbi do an excellent job in this book outlining exactly that: how to apply Google Classroom to real teaching problems.

Teachers are truly special people. Teachers like Alice and Libbi, who take the time to share their experiences, ideas, and solutions with other teachers, are truly a gift.

Jonathan Rochelle
Director of Product Management at Google

Introduction

A few short decades ago, the idea of every student having access to a computer in the classroom seemed like a science-fiction fantasy. How times and technology have changed! Thousands of schools today provide every student with a digital device (computer, laptop, tablet, etc.). As more schools move toward this one-to-one environment, teachers must respond to and lead the cultural and technological shift that's occurring in the classroom. Unlike the students of this digital generation, however, relatively few of today's teachers grew up with computers in their schooling. It's no wonder that one of the questions we hear most frequently from teachers is: "How can I effectively implement digital tools in my classroom when I don't fully understand them myself?"

Handing students a device can seem scary. But, as educators, we have to continually reexamine, reflect on, and adapt our teaching practices to suit our students' needs. Adding technology to our classrooms isn't optional; it's a must if we're to equip our students for their futures. We know this. And yet, we also know it's challenging because moving to a digital platform is more involved than simply shifting what you are currently doing into a digital form.

One resource we've found to be extremely useful for teachers and their students is Google Apps for Education (GAfE). In fact, millions of students have been introduced to digital tools through GAfE. Designed for true collaboration, GAfE allows for authentic and shared group work. For example, with Google Docs, an app in GAfE, students and teachers are able to work together on the same document at the same time. The cloud-based nature of Google Docs makes it easy for students to share and publish their work with an authentic audience, which brings more meaning than working solely for their teacher. In short, Google Docs creates a better, more engaging learning environment by empowering students to easily produce content and receive peer, teacher, and community feedback faster than ever.

While Google Docs and Apps allowed for new things to happen in the classroom, they were not designed expressly for the teacher and students. These digital tools helped, but left teachers with the need to create sometimes elaborate systems to fit them into the flow of their classrooms. Google Classroom was designed to simplify the use of Google Docs and Apps by teachers and students. Google Classroom was officially included in the Google Apps for Education suite in August 2014. Its popularity exploded as it fulfilled a real need for teachers, students, and classrooms. An online interface for distributing and collecting digital work, Google Classroom makes it easy for teachers to facilitate a digital or blended learning classroom. Designed from the ground up with teacher input, the platform allows teachers to connect with students, share innovative and technologically rich resources, and build creative projects and instructional elements into their daily lessons. Google Classroom streamlines the process for getting students on the same page, communicating with others, and sharing ideas through collaborative projects.

As with any new tool, the question is: What can I do with it? Shortly after Google Classroom came out, Alice Keeler posted a blog post on "20 things You Can Do With Google Classroom" in answer to that question. The post's popularity led us to expand on the ideas and ultimately write *50 Things You Can Do With Google Classroom*. In this book, you'll find practical ideas for using Google Classroom, along with detailed instructions and screenshots to guide you in the learning process. We're excited to show you how to use this powerful resource in your classroom. Let's get started with an overview of Google Classroom.

Introduction to Google Classroom

Google Classroom is an online platform that allows teachers to streamline the process of going digital with their students. Teachers are able to create a class within Google Classroom, provide their students with an add code or invite them via email, and quickly start communicating with students about class information, assignments, and documents. Part of the Google Apps for Education suite,

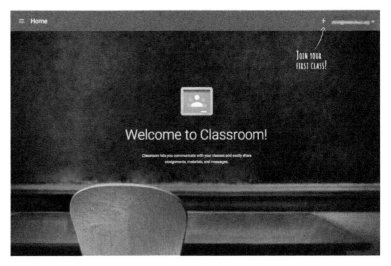

Google Classroom integrates the other Google Apps to make for a more seamless educational experience. Google Classroom is only available to schools with Google Apps for Education or Google Apps for Non-Profit accounts.

Google Apps for Education (GAfE) is a free suite of Google tools available to schools. GAfE includes Gmail, Google Calendar, Google Drive, Google Docs, Google Sites, YouTube, Google Classroom, and more. Schools can sign up at https://www. google.com/work/apps/education/. Schools with Google Apps will have a Google Apps domain manager who can assign accounts to staff and students. The Apps manager can also enable or disable features and Google Apps products, depending on the needs of the school. By default, Google Classroom is enabled for teachers and students and thus does not need to be enabled by the Google Apps domain manager from the school or district.

Overview of Google Drive and Docs

Google Drive (http://drive.google.com) provides users with online storage for digital documents. Additionally, Google Docs for text documents, Google Slides for presentations, Google Sheets for spreadsheets, Google Drawing and Google Forms can be created within Google Drive. Students and teachers are able to utilize these productivity tools to create documents from the cloud. This means that no hardware needs to be installed on student devices; only an Internet connection is needed. Students and teachers are able to access the documents in Google Drive from any Internet-enabled device, including mobile devices. Students are freed up from having to be on a particular device to do their work.

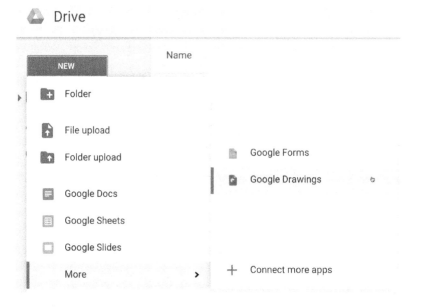

One of the most powerful features of Google Documents is real time collaboration. Students and teachers are able to edit concurrently on a document. Documents are accessed through a shared web link or through Google Drive. Collaborators of the same document simply need to open the document to find the current version and to work side by side with other users. As this collaboration is all cloud-based, collaborators do not need to be in the same room. Some classrooms are having students collaborate globally with students in other countries. This ability eliminates version confusion. Collaborators on a document have the confidence of knowing they are looking at the current version. Real-time collaboration has facilitated an easier process for peer review and for teachers providing feedback.

On Google Docs, Slides, Sheets, and Drawings, it's possible to insert comments without editing the document directly. Collaborators are notified of these comments via email or through the comments thread located via the "Comments" button in the documents. This aids in the feedback process and helps students to reflect on and evaluate their own work. Because the document is accessed through a link rather than an emailed attachment, students do not have to stop working on their document to wait for comments on their work. Within text documents, Suggesting mode allows for suggested editing by the teacher or peers.

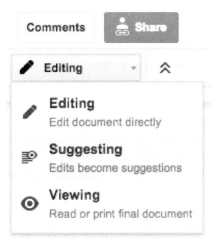

Using Google Drive and Google Docs has enabled many classrooms to use digital tools and reap the benefits of real-time collaboration and commenting. Google Classroom goes a step further to provide a way to facilitate technology integration into a traditional, online, or blended class. In Google Classroom, teachers can easily share assignments, documents and resources with students in an environment that lends itself to collaboration and creativity. Google Classroom provides Google Drive management and digital classroom interaction for schools using Google Apps for Education (GAfE). Teachers can use Google Classroom to post announcements and assignments to their classes. Students are able to turn in digital work directly through Google Classroom.

Getting Started

To start using Google Classroom, the teacher visits http://classroom.google.com or locates the Google Classroom icon in the Apps chooser. In the Apps chooser, Google Classroom will be located under the "More" apps option.

The first time a teacher enters Google Classroom, he will be prompted to indicate if he is a teacher or a student. If the teacher accidentally chooses student, he will need to contact the GAfE manager to allow teacher access. To get started, the teacher will locate, in the upper right hand corner of Google Classroom, a plus button (+) to allow the teacher to "Join class" or "Create class."

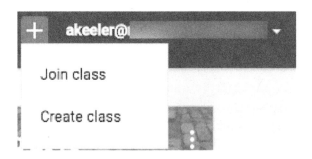

Choosing "Create class" creates a tile in Google Classroom and sets up a folder in the teacher's Google Drive. A tile is displayed on the Home screen for each class the teacher creates. For teachers who have multiple sections of a class, they will need to create a class for each section of the course. Alternatively, the teacher could choose to have all of the students from the different sections enroll in the same class.

Google Classroom automatically creates a "Classroom" folder in Google Drive for both the teacher and the student. Nested inside this folder are the folders for each specific class that is created. Nested within each class folder is a folder for every assignment the teacher creates in Google Classroom.

From the Google Classroom home, clicking on the class tile opens up the course. The class stream is now visible, and teachers can get started by adding announcements or assignments. The class code for students to join is located on the left hand side.

Students can join the class by going to Google Classroom, choosing "Join class" from the plus button in the upper right hand corner, and entering the class code. Students do not have the option to "Create class." Alternatively, students can be invited to a class from the "Students" tab in Google Classroom.

Google Classroom is a closed environment. Only students who have joined or were invited to join Google Classroom are able to view the assignments and announcements. This allows for students' names and comments to remain private.

Teacher View:
A Quick Tour

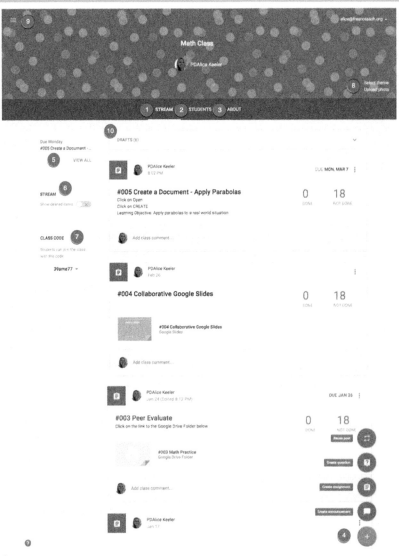

(1) The Stream

After clicking on a Google Classroom tile, the stream will be visible. The stream is the list of announcements and assignments. The teacher's view allows the teacher to create an assignment or announcement right from the stream. Students are able to post comments to the stream that are visible to the entire class.

(2) Students

Google Classroom shows a list of the students who are enrolled in the class. Teachers can invite students to Google Classroom by clicking on the "Students" tab. In this tab, the teacher can also set the ability for students to comment in the stream. Teachers can also remove students from the class from the "Students" tab. An option to mute a student from being able to post in the class stream is available there as well.

(3) About

The "About" tab allows the teacher to list information about the class, such as classroom location or course time. In the tab, the teacher can also attach files or links to important resources, such as the syllabus or the link to the classroom website.

(4) Create an Assignment

At the bottom right corner is a plus icon that allows the teacher to create an announcement, assignment, ask a question, or reuse a post from a previous class or the current class. From the stream, the teacher can edit, delete, or move an assignment up in the stream by clicking on the three dots in the upper right corner.

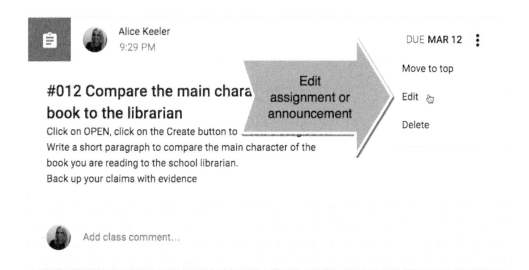

(5) Upcoming Assignments

Work that has not yet reached the due date are listed on the left side of the stream. The full list of assignments can be found under the menu button on the upper left side.

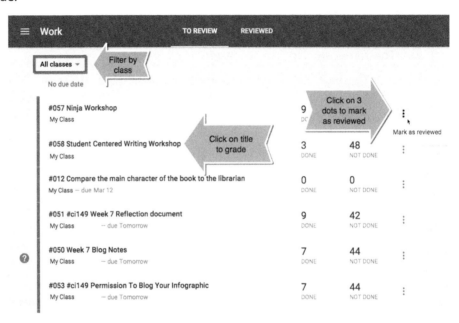

(6) Show Deleted Comments

The teacher is able to monitor student comments in the stream by deleting comments that do not belong. Deleted comments are removed from the student view yet are still accessible to the teacher. Toggling "Show deleted items" allows the teacher to view deleted comments.

(7) Class Code

Students go to http://classroom.google.com and add the class code after choosing to "Join class." The class code can be reset or disabled to stop students from joining the Classroom.

(8) Change Class Theme

Teachers can change the class theme by clicking on "Change class theme" within the Classroom header. This will change the header image and color theme of the Classroom.

(9) Menu

The three lines in the upper left corner allow the teacher to return to the home page, access the list of assignments from all classes, and toggle between classes.

Google Classroom settings can be found at the bottom of the menu bar. Teachers can choose to turn off email notifications for themselves in the settings.

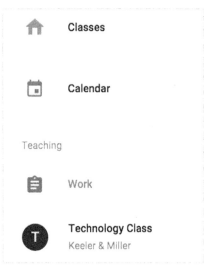

(10) Draft Assignments

Posts can be saved to publish later. Co-teachers are able to review and edit assignments before publishing. Click on drafts to view posts, edit, and then publish to the stream.

Assignment Screen

When the teacher links to the assignment, either from the menu by choosing "Assignments" or by clicking on the assignment title in the stream, the assignment screen displays a list of students and their submissions. Clicking on any student's name displays the student's work submitted and provides the opportunity to leave a private comment for the student.

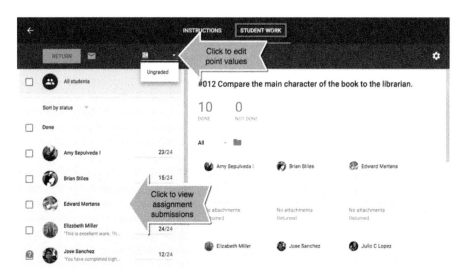

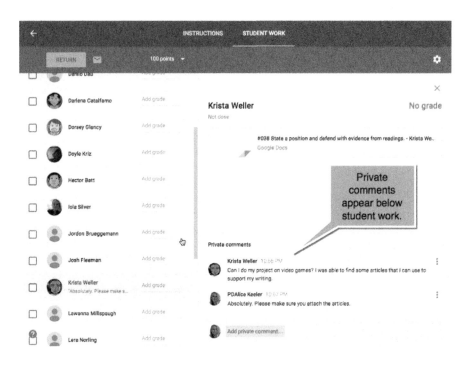

Student View:
A Quick Tour

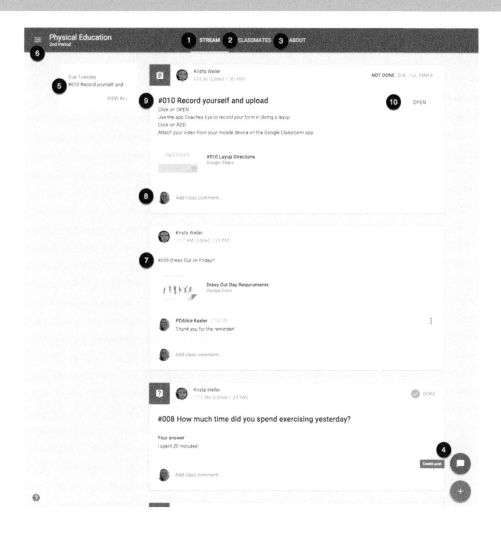

(1) Stream

The stream is where students can view assignments and announcements posted to Google Classroom.

(2) Classmates

From the "Classmates" tab, students can view a list of the other students enrolled in the Google Classroom.

(3) About

The "About" tab shows any information the teacher has posted about the class. Resources, such as the syllabus and other documents pertaining to the class but are not assignments, may be located in the About tab for students to locate easily.

(4) Share with Your Class

At the top of the stream, the students can post a global comment or question to the class. Students can attach files, links, or Google Drive documents to the stream in their comment. Teachers can mute a student's ability to do this from the "Students" tab.

(5) Assignments

The left hand side of the screen contains a block with a list of assignments that are due soon: within 6 days. The full list of assignments is also accessed by clicking on "View All" in the assignment block. This list is located via the menu icon in the upper left hand corner.

(6) Menu

Students can switch classes, return to the main menu, or view a list of their assignments through the menu icon in the upper left corner.

(7) Announcements

Documents posted by the teacher as announcements are created as "view only" for students. Students can read the announcement and open any attached files or links in the announcement.

(8) Add Comment

Unless muted by the teacher, students can post a response or question to any announcement or assignment. Students are not able to attach files or links to the announcements or assignments.

(9) Assignment

Students can find assignments the teacher has posted in the stream and then access links or attachments in the assignment directly from the stream. Assignment template documents provided by the teacher are not visible in the stream. Templates are found on the assignment submission screen.

(10) Open

For each assignment, an "Open" button is available. Students click on the "Open" button to reveal the assignment submission screen.

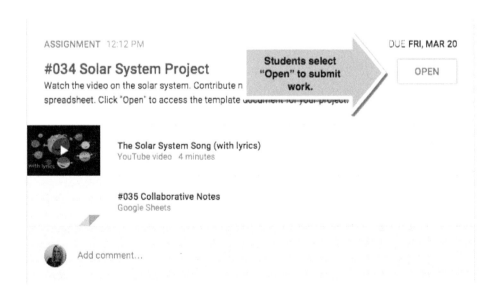

Assignment Submission Screen

Google Classroom allows students to submit digital work. Clicking on the "Open" button from the assignment takes students to the assignment submission screen. If the teacher has chosen to provide a template for part of the assignment, the template documents distributed to the students are available. Students click on the document title to open the document and edit. Students can add additional files and turn in their work through the assignment submission screen.

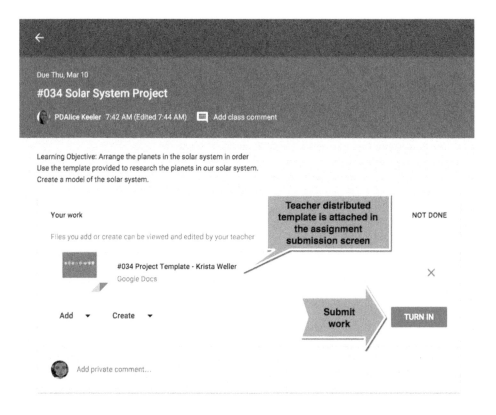

Google Classroom App

Google Classroom has a free app for both the Android and iOS in the App Store. The app allows students to view the class stream and submit their work. In the app, students can access the stream even when offline. The Google Classroom app also allows students to take pictures of their work through the app and submit the images as an assignment.

The Google Classroom app allows sharing from other apps, including Google Docs, Google Slides, and Google Sheets. Work completed in these different apps can be submitted to Google Classroom from the other app. This allows students to utilize a variety of apps to express their ideas.

50 Things You Can Do With Google Classroom

1. Make Class Announcements

Google Classroom gives teachers a place to post their class announcements. Teachers can post announcements to the stream and send an email to the students' Gmail accounts. Students are able to locate older announcements by scrolling down in the stream. Unlike verbal announcements or those written on a whiteboard, announcements in Google Classroom are accessible outside of class. The Google Classroom also allows students to make comments on the announcement. This transforms what might traditionally be one-way communication into two-way communication.

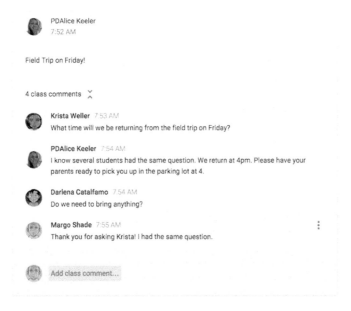

2. Share Resources

Google Classroom allows teachers to take a document, video, or link and push it out to their students. Utilizing Google Classroom as the consistent location for students to obtain digital resources maximizes classroom instruction time. When students are not being directed to multiple locations to find resources, the flow of the classroom is improved.

To share a resource, click on "Announcement" at the top of the Google Classroom stream. There are four icons along the bottom of the announcement creation box.

- The paperclip icon attaches files that are saved on the computer.
- The Google Drive icon allows Google Docs or other files stored in the teacher's Google Drive to be attached to the announcement. If the file in the Google Drive is shared privately, the sharing settings are modified to allow students in the class to view the file without additional steps by the teacher.
- The YouTube icon allows the teacher to paste the URL of a YouTube video they already have located. Alternatively, Google Classroom provides a search box to locate YouTube videos.
- The fourth icon allows the teacher to paste the URL of an Internet resource.

Documents that are shared through the Announcements tool in Google Classroom are shared with the students as view-only files. These resources are not viewable to students not in the class, unless the teacher has also shared them in other locations besides Google Classroom.

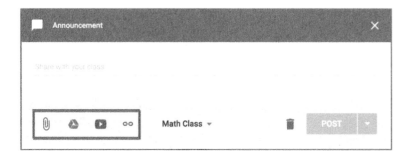

3. Keep Multiple Files in an Assignment

Google Classroom allows for multiple attachments to a single assignment. Teachers can assign the students multi-phase projects and provide a template for each of the different phases.

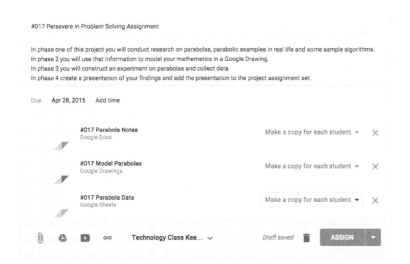

4. Create a Lesson

More than simply assigning work to students, Google Classroom allows the teacher to build a complete lesson. At the top of the stream in the Classroom, the teacher can click on "Assignment" to start building a lesson set. The description area of the assignment allows the teacher to provide directions to students for completing the lesson and assignment.

Students are able to move through the lesson more efficiently when resources and materials are presented in a logical order. Creating an instructional document in Google Drive and attaching it using the Google Drive icon can start the lesson set.

To supplement the instructional document, the teacher can attach YouTube videos or videos located in the teacher's Google Drive as the next part of the lesson set. Engaging videos, screencasts, or short instructional pieces can provide additional understanding beyond the instructional text. Videos can also be utilized to differentiate instructions for students. Providing multiple videos that address different learning modalities or ability levels can help students choose an instructional option that works for them.

Websites that allow students to practice some of the skills in the lesson can be provided as part of the lesson set. Collaborative documents may also be added to the lesson set to allow students to brainstorm or crowd-source information. Typically, the last part of the lesson set is the assignment for the student to complete. Attaching a graphic organizer or template document provides the task for the student to complete.

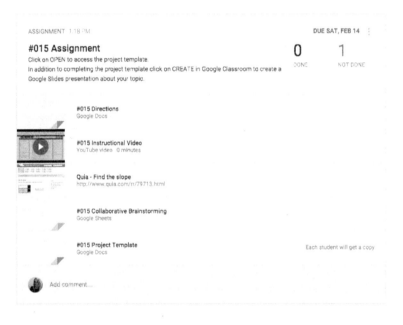

5. Go Paperless

Passing out and collecting papers in class can take up valuable instructional minutes. Using Google Docs and Google Drive allows the teacher to go paperless. Teachers can require students to create a Google Document from scratch right in Google Classroom, or the teacher can provide a template for the students to fill out. Google Classroom can create a copy for each student, giving them turn-in buttons for when they are done.

6. Easily View Student Submission

Google Classroom counts how many students have and have not submitted an assignment. Teachers can find the number of student submissions clearly displayed in the upper right side of each assignment in the stream. Click on the number done to see a list of students who have submitted along with their assignment submission.

7. Simplify the Turn-In Process

When using Google Documents, students often forget to change the sharing settings or to turn in their work. As a result, the teacher cannot view or access the assignment. Google Classroom eliminates this issue by placing the document in the teacher's and the student's Google Drive immediately. This gives both of them access to the document.

Google Classroom provides the students with a turn-in button to signal the teacher they are ready to have their work assessed. When students use the turn-in button to submit an assignment, the ownership of the document is transferred to the teacher. After submitting, the student still has viewing rights, but they are unable to edit or change the content.

8. Protect Privacy

The documents the students submit in Google Classroom are shared only between the teacher and the student. Students do not have access to the work of other students. Google Classroom places all of the students' documents in a single folder in Google Drive; however, the full contents of the folder are not shared with all of the other students.

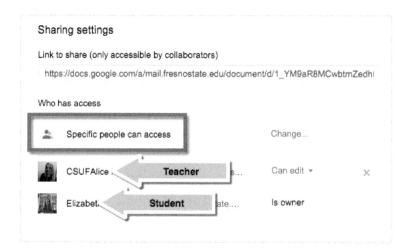

9. Encourage Classroom Collaboration

When creating an assignment, the teacher is able to choose whether documents are shared as View Only or that documents are shared so students can edit. By choosing "Students can edit file," all students in the class are able to edit the same document at the same time. This allows every student to contribute to a class project or activity.

Google Slides and Google Sheets are great for simultaneous class collaboration. On a Google Slides presentation, each student can work on an individual slide. Google Sheets allows each student to write in a separate cell. This is useful for crowdsourcing data and information. Google Sheets also lets students work on their own individual tabs. The teacher is able to review all of the students' work in one document.

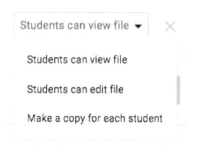

10. Reduce Cheating

Students do not have access to the Google Classroom assignment folder. The assignment folder is accessible by the teacher only. Since the class documents are not in a *shared* folder, the students do not have the ability to copy other students' work from the folder. This is an upgrade from simply using Google Drive; sharing a Google Drive folder with the class allows for all students to view the work of other students, which, potentially, can lead to copying.

Within each student Google Document, the teacher can use the revision history to see when the document was edited and who did the editing. Revisions are frequently recorded. When a student has only one revision, this could signal that the student copied the document of another student.

11. Create a Discussion

A Google Sheets spreadsheet can be utilized to gather student ideas on a discussion topic. Discussion questions can be added on an individual page, and additional tabs can be used for multiple questions. The Google Sheet can be shared with students for editing access, and each student can answer the discussion questions. This allows for all students to have a voice in the discussion even when they might struggle with speaking up in class. The spreadsheet discussion also allows the students to view their classmates' ideas in order to compare and contrast the ideas with their own. Student responses can be dragged around the sheet in order to articulate patterns and differences in student opinions.

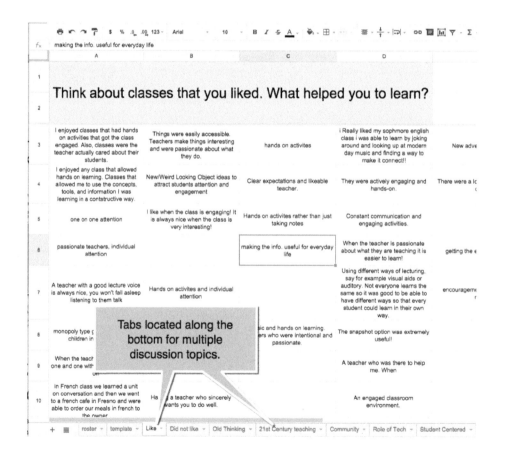

12. Organize Assignments with Due Dates

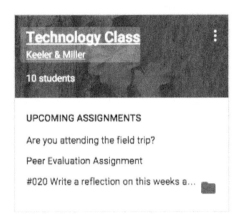

When creating an assignment in Google Classroom, the teacher is able to assign a due date that is clear for both the teacher and the students. Google Classroom clearly displays assignments that are not yet due on the Classroom tile for the student to see immediately after logging in.

Each assignment in the stream clearly indicates the due date in the upper right corner of each assignment description.

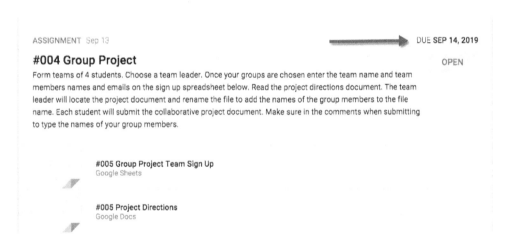

Overdue assignments are labeled as "Late" for the student. This notification clearly identifies missing or incomplete assignments in the stream for the students. Late work that is submitted is flagged as such for the teacher.

13. Give Feedback Before Students Submit

Providing students feedback while they are working on an assignment increases student motivation and is more instructive than providing feedback after the assignment is completed. It shifts the focus from the product to the learning process. In Google Classroom, the teacher can provide feedback by inserting comments into a student's Google Document before the due date. The teacher shifts from being an evaluator to being a coach or facilitator of learning.

In the teacher's view, Google Classroom creates an assignment-specific folder that contains each student's document. The assignment folder can be accessed from Google Classroom through the "Folder" icon that links the teacher to the assignment folder. The folder can also be accessed directly from Google Drive. While students are creating and editing the assignment, the teacher can open the assignment folder and insert comments. Feedback can be provided simultaneously as the students are working on the documents.

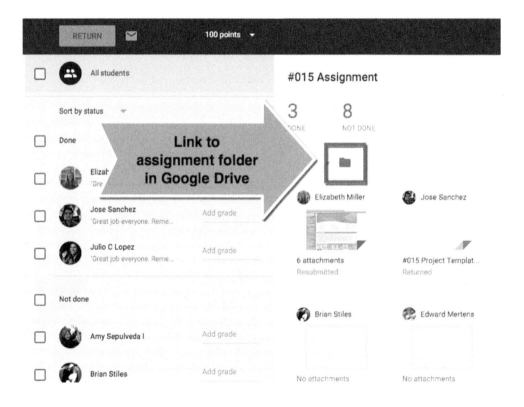

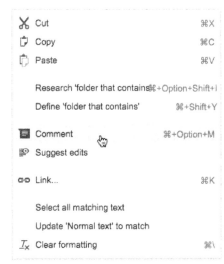

There are multiple ways to insert comments into a Google Document. The teacher can place the cursor at the point in the student's document where they wish to make a comment, or the teacher can highlight a section of text and right-click to choose "Comment" from the menu options. Teachers can speed up this process by learning the keyboard shortcuts for inserting comments. Control + Alt + M on a PC or Command + Option + M on a Mac will insert a comment. On a PC, holding down the Control key and pressing the Enter key will close the comment. Mac users use Command + Enter to close the comment.

14. Email Students

Sending email to students is simple through Google Classroom. Teachers can email students from any assignment page or go to the students' pages to send emails to selected individual students or all students.

15. Notify Students Who May Need Help

When viewing the stream in Google Classroom, teachers can clearly see the number of students who have completed the assignment and the number who have not. Clicking on the number of students who have not submitted the assignment displays a list of student names whose assignments have not been submitted. Google Classroom allows the teacher to send a bulk email to all students who have not submitted the assignment. In the email notification, the teacher can increase success by encouraging the student to complete the assignment.

DUE NOV 20, 2014

53
DONE

80
NOT DONE

16. Assignment Q&A

When an assignment is posted to Google Classroom, the students have the ability to comment on it. No longer do students have to wait to be called on to ask a question. This extends learning beyond the classroom walls and allows students to ask questions at any time from any location. When the teacher posts a response, it is available to all students.

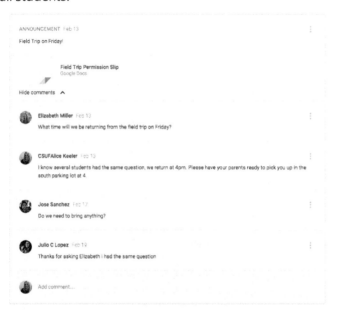

17. Create an Ad Hoc Playlist

Google Classroom allows the teacher to attach multiple YouTube videos to an announcement or assignment. Teachers can create a playlist of YouTube videos right in Google Classroom. If YouTube is blocked (as it may be in some schools), teachers can attach videos to the playlist that are stored in Google Drive. Students can access the video list right in Google Classroom without having to navigate to multiple websites.

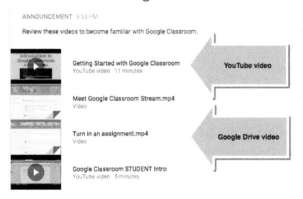

18. Email Feedback

Directly from Google Classroom, the teacher has two options for emailing feedback to students. The first option is for the teacher to send a global note to the entire class. This is done using the "RETURN" button. When choosing the "RETURN" button, a pop-up box opens where the teacher can provide a global note for all students who have submitted the assignment. This note is emailed to the student in their GAfE accounts. The global notes are also available in the private comments on the assignment page.

The second option allows for the teacher to send an email to an individual student or a selected group of students. Click on the email icon to launch the teacher's Gmail to compose an email to the students.

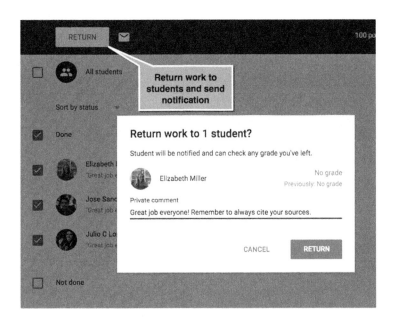

19. Create Folders

Google Classroom creates folders in Google Drive. What once was a cumbersome process in Google Drive is now automatic. The teacher and each student will find a "Classroom" folder in Google Drive. In the "Classroom" folder, the teacher has a folder for each assignment. Student work can be accessed from Google Classroom or directly from the folders in Google Drive.

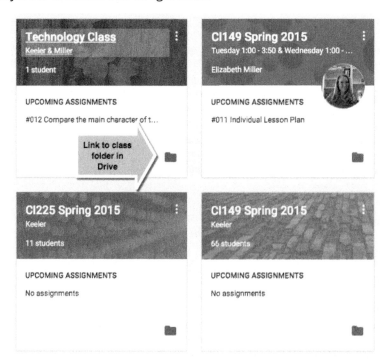

To make access easy, Google Classroom also provides links to the Google Drive folders. A link to the class folder is located on the class tile on the Home page. In the grading view of each assignment in Google Classroom is a link to the specific assignment folder in Google Drive.

20. Link Directly to Student Work

When students submit assignments, their documents are available through a link in Google Classroom. The assignment page contains a list of all student work. When teachers click on a submission, they are linked directly to the students' documents without having to search for them in Google Drive. This allows teachers to provide feedback to students quickly and efficiently.

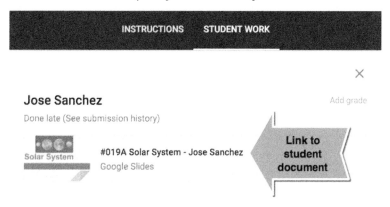

21. Collect Data

There are several ways to use Google Classroom to collect data. The teacher can share a Google Form with students as an announcement in the stream. The teacher can also create an assignment and attach a Google Sheet as "Students can edit." This allows students to organize their information directly on the spreadsheet. Asking students to crowd-source resources onto a single document is another quick way to collect data from students. This also allows students to start collaborating right away.

22. Share with Multiple Classes

If a teacher teaches multiple sections of the same course, Google Classroom can create the assignment in each section. From the stream, teachers can choose to create an assignment or an announcement. After creating an assignment title and description, the teacher can set the due date, attach files or links, and can copy the assignment to multiple class sections. The list of class sections is available along the bottom of the assignment or announcement in a dropdown menu. Assignments can be copied for up to 10 classes.

23. Collaborative Note-Taking

Collaborative note-taking is a great way for students to have a more complete set of class notes. The collaborative aspect allows students to work together so they do not miss information from the class. The teacher can create a Google document and share it as "Students can edit file" in an assignment. Then the teacher can designate specific students to be note takers for a discussion or activity. The students can collaboratively take notes on the document, and those notes are easily accessible by the other students through the announcement in Classroom.

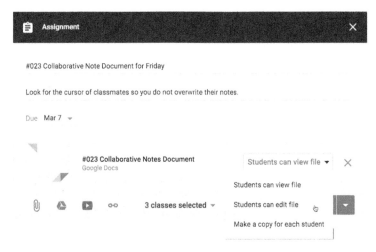

24. Display Student Work

When students submit work in Google Classroom, their work is saved in a folder in Google Drive. The teacher can attach files by creating a new announcement and clicking the Google Drive option. With a student's permission, the teacher can share an announcement with a link to the student's work that is available in his/her Classroom Google Drive folder.

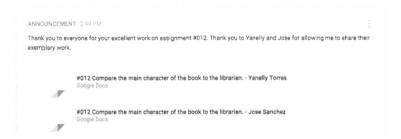

25. One Student, One Slide

Students can collaborate through Google Slide presentations. The teacher can create an assignment that includes a Google Slide presentation set as "Students can edit file." The teacher can modify the slide master to provide a template for student work. The students can access the presentation through Classroom and insert their own slides.

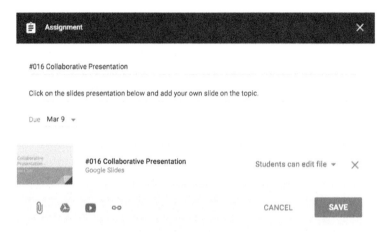

26. Target Parent Phone Calls

Google Classroom shows which students did not complete an assignment. By clicking on the number above "NOT DONE," the teacher is provided with a list of students whose assignments have not been submitted. The teacher can use this list to make parent phone calls or send emails.

27. Polling

An assignment can be created to find out which students are attending a school event. If attending the event, the student will click on "MARK AS DONE." If not attending, the student does not respond in any way. The teacher will now have a clear list of which students are attending. Unlike a Google Form, the teacher now has a clear list of who is not attending through the "NOT DONE" list.

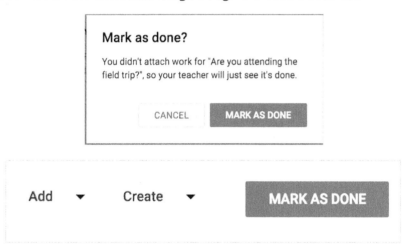

28. Share a Document with the Class

Distributing documents can be a consuming task for teachers. Teachers spend a lot of time printing documents, going to the copy machine, passing them out in class, and then waiting on students to locate papers that they were already given. Google Classroom makes document distribution simple. Attaching a document, either through an announcement or within an assignment, allows students to quickly locate the class materials and return to the resource when they need it later.

29. Know Who Edits a Collaborative Document

Instead of sharing a Google Doc as "Anyone with the link can edit," Google Classroom allows the teacher to limit editing privileges to the students enrolled in the class. When the students are signed into Classroom and have accessed the document, their individual icon appears at the top of the page. This replaces the anonymous animal. The teacher can also see where a student is working in the document because a cursor appears with the student's name. The teacher can use revision history to identify who edited the document and see when the changes were made.

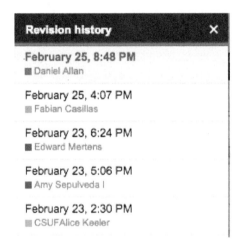

30. Link to a Website

Relying on students to type in a web address correctly also costs teachers instructional minutes. It is easy to make mistakes when typing a URL, and it becomes a challenge to get every student linked to the same website in a reasonable amount of time. In Google Classroom, the teacher can provide links to websites. The students can click on the link and get to the page easily and quickly.

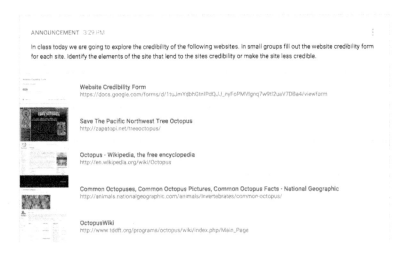

31. Peer Feedback

Students can provide feedback to their peers using Google Slides in the Google Classroom. A teacher can create and share a Google Slides presentation with the students using the "Students can edit file" permission setting. Then each student can create a slide with his/her information. The other students can view their work and insert comments on their classmate's slide to provide feedback.

32. After-Hours Help

Students sometimes struggle to complete homework or projects on their own after school hours. In Google Classroom, students can post their questions in the stream to receive a peer or teacher response. Comments or questions can also be added below each assignment or announcement and can be accessed at any time by students or teacher. As these are posted in the stream, everyone in the class can read or respond to the comments. This way, the other students benefit from the questions and responses.

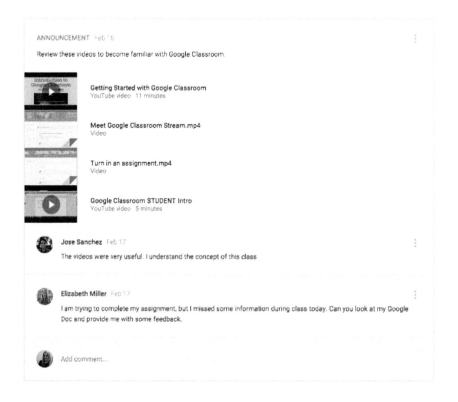

33. Distribute Notes

Using a Google Doc in Google Classroom allows students to focus on classroom activities and discussions instead of note-taking. The teacher can post the notes from the lesson in the classroom as an announcement. The students can quickly access the notes so they can spend their class time interacting and talking about their ideas.

34. Sharing Informal Learning

As students discover ways to connect their classroom learning to their lives outside of school, they are able to share their experiences in the Google Classroom. Students have the ability to share pictures, Google Docs, YouTube videos, and other links with the class. Other students and the teacher can make comments and reply to the student. This encourages students to engage in sharing and discussing their informal learning.

35. Email the Teacher

Clicking on the About tab shows the student who teaches the class and an icon to email the teacher. Clicking on the icon launches a new message window. Because the email will come from the student's GAfE account, the teacher can be assured the message is from that student.

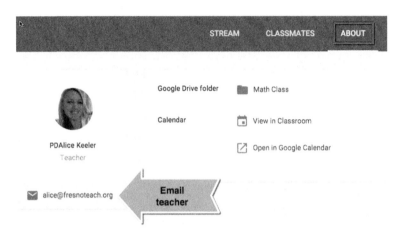

36. Eliminate Schlepping Papers Home

Google Classroom eliminates the need for teachers to carry papers home for grading; students can submit all of their work in a digital format. This eliminates collecting, organizing, and passing back papers. For assignments that are not digital, the students can use the Google Classroom app to take pictures of their physical work and turn it in digitally. Students who are using Chromebooks or other laptops can insert pictures by using the snapshot feature in Google Docs. This turns a physical paper into a digital document.

37. Student Projects

Google Classroom allows students to attach multiple artifacts when submitting their assignments. In one place, the students can submit each element of their projects. Students may include a variety of attachments, such as images (like screenshots), links to web projects, notes, citations or references, Google Slides presentations, and the final product. Students cannot submit Google Drive documents or presentations that they do not own. Google Classroom neatly organizes these elements and easily allows the teacher to assess students' progress.

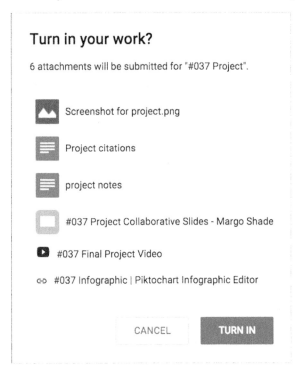

Turn in your work?

6 attachments will be submitted for "#037 Project".

- Screenshot for project.png
- Project citations
- project notes
- #037 Project Collaborative Slides - Margo Shade
- #037 Final Project Video
- #037 Infographic | Piktochart Infographic Editor

CANCEL   **TURN IN**

38. Have One Place for All Files

Google Classroom supports automatic Google Drive management and organization. Any documents students submit via Google Classroom are saved in Google Drive in an automatic file system. This gives the teacher one central location to check for student work. When students do work in other programs, they are able to take a screenshot and submit the screenshot to Google Classroom. Students choose "Upload File" on the assignment submission page to turn in screenshots. These screenshots are also saved to the Google Classroom assignment folder in Google Drive.

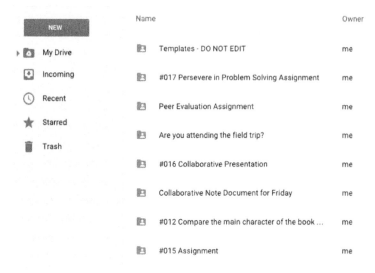

39. Document Digital Work

Teachers can create an assignment in Google Classroom and have students provide the link to their non-Google digital work. On the assignment submission page, the students have the option to turn in a URL. The students can turn in websites, wikis, or other digital resources by linking to them in Google Classroom.

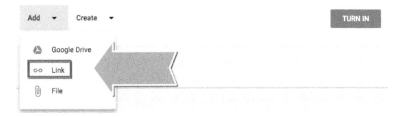

40. Students Create Google Docs

When viewing the assignment submission page, students can click "Create," which allows them to start a new Google Document, Presentation, Spreadsheet, or Drawing. This file is automatically attached in Google Classroom and titled the same as the assignment. The document title is appended with the student name and saved in the assignment folder in Google Drive.

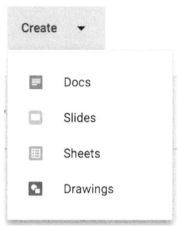

41. Clearly Identify Student Work

When a document is shared with students as "Each student gets a copy," the new document shares the title of the template document and the student's name is appended to the document title. When the teacher looks in the Google Drive assignment folder, it is easy to identify which document belongs to which student. This cuts down on the issues surrounding papers and assignments without names.

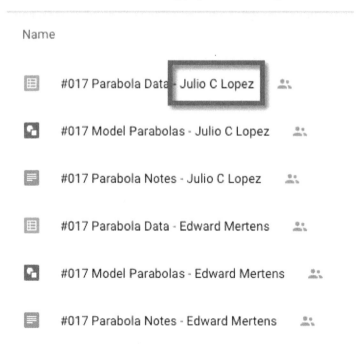

42. View Assignments

Google Classroom provides the teacher and students a list of assignments that have been previously assigned. Locate the list of assignments under the menu icon in the upper left corner.

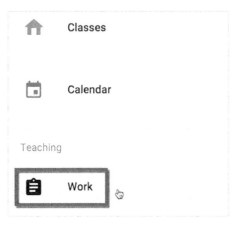

The teacher's list is split to show assignments "To Review" and assignments that are already "Reviewed." This makes it easy for the teacher to score and evaluate student work.

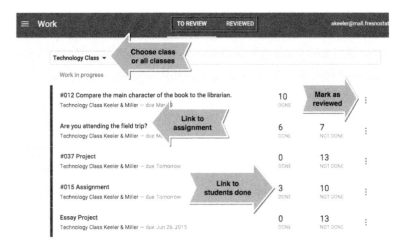

In the student view, students are able to find all the assignments that the teacher has posted. By clicking on "View All," the student sees assignments, which are separated by ones "To-Do" and "Already Done."

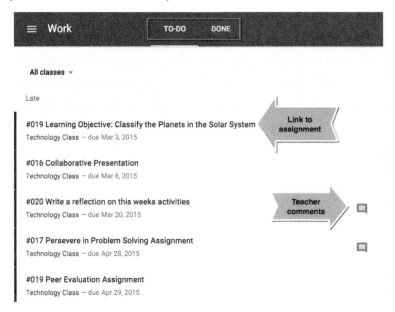

43. Collaborate with Peers (PLCs)

Teachers can join a classroom as a student by clicking on the plus (+) button in the upper right corner of the home screen. Teachers choose "Join class" and enter the class code of the PLC Classroom that was set up. This allows a grade level or subject area team to create a Google Classroom for the teacher group. Meeting notes, data, and other documents can be linked and shared from Google Classroom. Teachers can also submit their classroom results from benchmark tests or other projects by using the "TURN IN" button in Classroom.

In the same way, teachers can join a Google Classroom as a student for Professional Development. Professional Development activities and resources can be organized in the Google Classroom stream. Teachers can then easily access the resources from the Professional Development at any time through the class tile in Google Classroom.

44. Virtual Office Hours

The "About" page in Google Classroom allows the teacher to post links and resources for students to access throughout the school year.

Copying the permalink to a Google Hangout and linking to it in the "About" page of Google Classroom allows the teacher to provide Virtual Office Hours to students.

Students can easily join a chat with their teacher during office hours by clicking on the link on the "About" page.

45. Virtual Faculty Meetings

Meetings can take up a good amount of a teacher's already busy schedule. Google Classroom makes it possible to reduce the number of faculty meetings. The school administrator can ask the faculty to join a Google Classroom. Short videos can be linked in Classroom to facilitate a flipped approach to faculty meetings. Links to Google Forms can be provided to have teachers provide data or respond to polling questions. Different departments can post announcements to the stream to share their news.

46. Streamline Counseling

High school counselors can invite all the students on their caseload to join Google Classroom. This allows counselors to consistently connect with their students and share important information. Rather than make announcements in the school bulletin, the counselor can create an announcement in Google Classroom and target the specific group of students who need the information. The counselor can also share resources about upcoming events or activities. Students can "MARK AS DONE" different tasks the counselor creates. This makes it easy to identify the students who are missing paperwork or did not complete an important task - such as an SAT application or a college essay. The counselor and students can also stay connected using the email option in Classroom.

47. Observe Another Classroom

Collaboration is one of the best things teachers can do to improve their practice. In Google Classroom, other teachers (or administrators) are able to enroll as students in other Google Classrooms. This is done by obtaining the class code from one's colleagues and joining their classes. This allows teachers to observe what type of assignments are being used and best practices for utilizing Google Classroom.

Note that teachers can only join the Google Classroom of another teacher in the same domain or one that has been whitelisted. This allows for new teachers to work with mentors and collaborate across classrooms. Adding a teacher as a co-teacher is another way to allow for observing another class. Co-teachers are able to be added to a Google Classroom using the "Invite teacher" button located in the about tab.

PDAlice Keeler

Teacher

✉ alice@fresnoteach.org

INVITE TEACHER

48. Watch Students Do Homework

Since Google Classroom is accessible to students on the Web, they are able to submit work at night and on weekends. Teachers can observe student work submitted 24/7. Student documents are accessible while students are working on them. Teachers can go into the assignment folder in Google Drive and see students working after school hours. This allows for real-time feedback and interactions.

49. Share Student Samples

Google Classroom assignments are saved to a Google Drive folder. This provides easy access to student work samples. Teachers can use the Google Drive icon when creating an announcement to find a previous assignment in the Google Classroom folder. Use the Control key to select multiple, student documents at once and share them through an announcement.

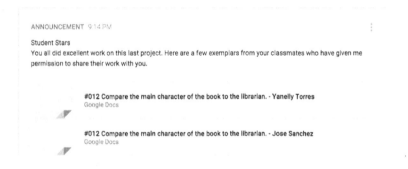

50. Provide Choices

Providing students with choices is not only a great way to differentiate instruction, but it also lets students have control in their learning. This leads to increased student engagement and motivation. In the directions of a Google Classroom assignment, the teacher can provide different options for students to demonstrate that learning objective. A suggestion is to label each choice with a name, such as A, B, C, etc. The teacher can then provide templates for the different choices in the assignment attachments. Students can click the X on the assignment submission page to remove templates they did not use.

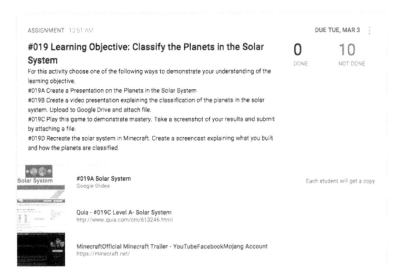

Google Classroom shows each student's attachments in a neatly organized list for the teacher to view. Each student is able to submit different and unique artifacts. Teachers can assess students' mastery of the learning objective easily by using the assignment page to view each student's submissions. Allowing students to demonstrate their learning in different ways, with one place to check student work, has never been easier for teachers.

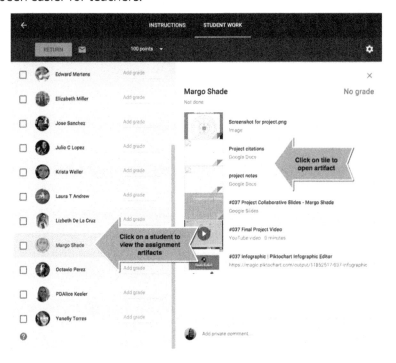

51. Reuse a Post

Google Classroom allows teachers to duplicate a post from another class, an archived class, or from the same class. A post is anything that has been added to the stream, including announcements, assignments or questions. The teacher can click on the plus icon and choose "Reuse post" to select a previously used post to add to the stream.

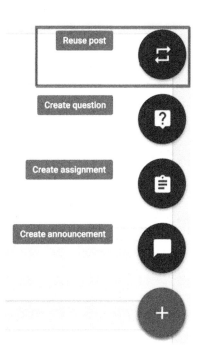

When reusing a post in Google Classroom, the teacher is able to update and edit the post before adding it to the current class. Reusing a post one at a time allows for thoughtful consideration of the updated the post and makes it relevant to the students in the current class.

Teachers must use caution when reusing a post and choosing the "Make copies of all attachments" checkbox. Google Drive is designed to eliminate redundant copies. Duplicating attachments creates multiple copies of a document with the same title. With rare exception, it is best to leave the "Make copies" checkbox unchecked.

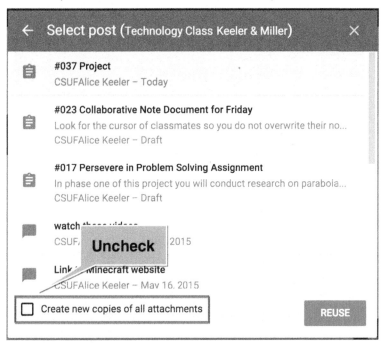

52. Use Google Forms

Google Forms allows for teachers to collect information from students and view all responses in one document. This saves the teacher from opening individual student documents. Viewing all submissions in one place also allows for a quick response to student needs. Instead of having students fill out PDFs of worksheets, teachers can have students submit their responses to assignments in Google Form.

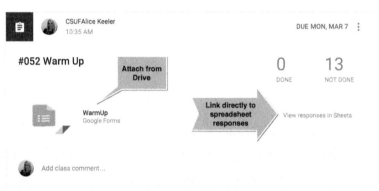

Teachers can create a Google Form in Google Drive and use the Drive icon on the Google Classroom assignment to add the Form to the stream. When no other attachments are added to the assignment, Google Classroom will automatically mark the assignment as done when students submit the Form.

Google Forms allows for quickly viewing a summary of the data from the "Responses" tab in the edit screen. The spreadsheet icon on the "Responses" tab on the Google Form allows the teacher to create or view the spreadsheet. After the spreadsheet is created, Google Classroom links to the spreadsheet right from the Form in the stream.

53. Create an Exit Ticket

Teachers can use the question feature in Google Classroom to create an exit ticket to use before students leave class. The teacher can choose the plus icon and select "Create question." The question acts as an assignment, and the teacher can attach files and links and assign points. If the question post has a due date, the question will appear on the Classroom calendar.

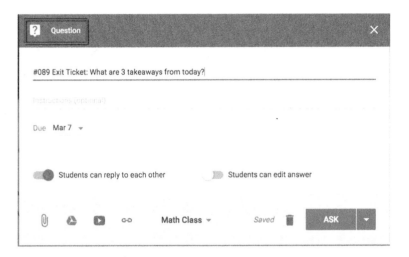

Google Classroom quickly counts and displays who has and has not answered the exit ticket question. Unlike paper or whiteboards, using Google Classroom allows the teacher to view all student responses at once. The teacher also has the option when creating the question to allow classmates to view one another's responses. Enabling this feature allows students to help with misconceptions and learn from the responses of others.

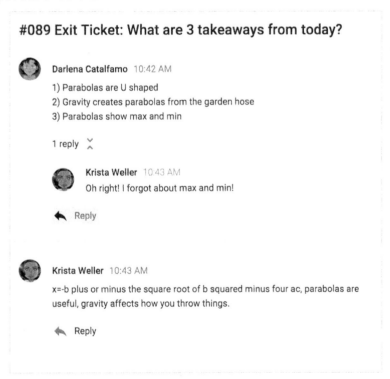

54. View Student Work

The assignment grading page displays assignment tiles on the right hand side. The tiles give the teacher a quick visual of who has and who has not attached work to the assignment. Clicking on the tile image launches the student's document quickly. This makes it easier for the teacher to provide the students with feedback before the due date. When multiple attachments are present, the teacher is taken to the student grading page to choose which attachments to view.

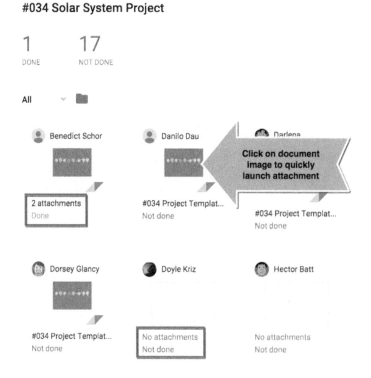

55. View Submission History

When a student submits an assignment, it is recorded in Google Classroom on the assignment grading page. Submission history shows a record of when assignments were submitted along with late, unsubmitted, and returned assignments.

Locating a specific student on the assignment grading screen displays the student's work, comments, and submission history. Submission history is located below the student's name. Clicking on "See submission history" shows the full record of submissions.

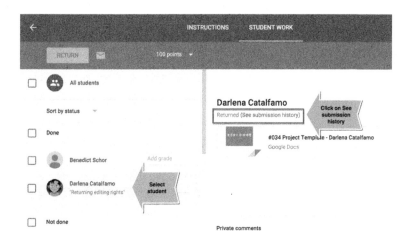

Darlena Catalfamo's submission history

Returned	Mar 6, 11:10 AM	PDAlice Keeler
Late unsubmitted	Mar 6, 11:10 AM	Darlena Catalfamo
Done	Mar 6, 11:10 AM	Darlena Catalfamo

CLOSE

Conclusion

So now you have read this book and have some ideas about using Google Classroom. Congratulations on taking strides to improve your students' education and experiences in your classroom! Understanding new technology and its capabilities is an excellent place to start. But we have to warn you: Using Google Classroom to distribute and collect worksheets will not improve your classroom or engage your students. Simply using Google Classroom will not affect significant change in your teaching or shift your pedagogical approach. If your classroom is teacher-centered, it won't matter what tools and technology you use; you'll get the same results either way. But when you combine the methods we've shared with strong, student-centered pedagogy and effective instructional strategies, your students will be more engaged in and excited about the learning process.

We've shown you how Google Classroom can help you do so much more, in terms of connecting students with resources, creating motivating and engaging instruction, and designing your classroom for collaboration. This book includes only fifty ideas for using Google Classroom, but there are many more ways this resource is being used. We hope you'll share your ideas with us and find new ones on Twitter by following the hashtag: #GoogleClassroom. You can also tweet: @alicekeeler or @millerlibbi to continue the conversation.

Part of effective teaching is constantly reflecting on your practice and making appropriate changes. Collaboration with other educators is an excellent place to begin. Host a coffeeEDU and connect with innovative educators in your area. (Learn more at www.coffeeEDU.org.) Seek new ways to engage your students and revolutionize your classroom. Read Dave Burgess's *Teach Like a PIRATE* and connect with innovative and engaging educators on Twitter, using the hashtag: #tlap.

What more can you do to revolutionize and technologically enrich the learning environment for your students? Continue to learn about and promote the use of educational technology. Become familiar with the SAMR model (Substitution,

Augmentation, Modification, Redefinition) of instructional technology usage. If you consistently integrate technology at the Substitution level, think about how you can move up to another level and begin to innovate and redefine what learning can look like for students. Share what you've learned and guide colleagues to implement technology to increase student engagement school-wide.

Above all, never let your teaching become static or routine. Continue to explore ways to improve and hone your teaching skills. As the saying goes: School's never out for the pro.

Additional tips and ideas for using Google Classroom can be found on Alice's blog: alicekeeler.com/classroom. Here are a few of her blog posts that may interest you:

- *A Tour of Google Classroom: http://goo.gl/dljD1l*

- *Google Classroom: Numbering Assignments: http://goo.gl/my6OWq*

- *Google Classroom: 6 Tips to Level Up: http://goo.gl/KTTWEl*

- *Google Classroom: Submit Screenshots: http://goo.gl/jVrKmr*

- *10 Things Students Want to Know About Google Classroom: http://goo.gl/lvXZOA*

- *Google Classroom: Feedback Faster with Chrome Extension Open Side by Side: http://goo.gl/Yj4NJb*

- *My Google Classroom Feedback Workflow: http://goo.gl/EllJT8*

- *Google Classroom: Create Group Documents: http://goo.gl/rpkirl*

- *Google Classroom: Provide Feedback Faster: http://goo.gl/RjORqg*

- *Google Classroom: 8 Steps Workflow to Create Files: http://goo.gl/ZMU2pu*

- *Google Classroom: 9 Tips for Attaching Google Drive Files: http://goo.gl/RKnwPO*

- *Google Classroom: Star Your Templates: http://goo.gl/Lxs7cz*

- *Google Classroom: Point Values: http://goo.gl/4yE6lf*

- *Google Classroom: Student Quick Sheet Guide: http://goo.gl/TbTLPv*

- *Google Classroom: OOPS, Fixing Assignments When You Forgot to Make Copies: http://goo.gl/uynu0s*

- *Google Classroom Deployment: Advice from David Malone: http://goo.gl/HtXsOk*

Acknowledgments

We would like to express our gratitude to the following people who have been instrumental in creating this book.

To Shelley (@burgess_shelley) and Dave Burgess (@burgessdave), for supporting this project from the beginning.

To Michael Montana, for his continual support and encouragement, along with all those home cooked meals.

To Jonathan Rochelle (@jrochelle), for his insights and input throughout the writing process.

To the educators who reviewed the text and shared their feedback from the lens of a classroom teacher: Erin Balfour (@MsBal4EVA), Krista Tsutsui (@kristatsutsui), Mike Mcsharry (@mikemcsharry).

A special thanks to those who participated in demonstrating features of Google Classroom: Julio Lopez, Edward Mertens, Laura Andrew, Lizbeth De La Cruz, Octavio Perez, Jose Sanchez, Amy Sepulveda, Brian Stiles, Yanelly Torres.

Also From

Dave Burgess
Consulting, Inc.

Teach Like a PIRATE
Increase Student Engagement, Boost Your Creativity, and Transform Your Life as an Educator
By Dave Burgess (@BurgessDave)

Teach Like a PIRATE is the *New York Times'* best-selling book that has sparked a worldwide educational revolution. It is part inspirational manifesto that ignites passion for the profession, and part practical road map filled with dynamic strategies to dramatically increase student engagement. Translated into multiple languages, its message resonates with educators who want to design outrageously creative lessons and transform school into a life-changing experience for students.

P is for PIRATE
Inspirational ABC's for Educators
By Dave and Shelley Burgess (@Burgess_Shelley)

Teaching is an adventure that stretches the imagination and calls for creativity every day! In *P is for Pirate*, husband and wife team, Dave and Shelley Burgess, encourage and inspire educators to make their classrooms fun and exciting places to learn. Tapping into years of personal experience and drawing on the insights of more than seventy educators, the authors offer a wealth of ideas for making learning and teaching more fulfilling than ever before.

140 Twitter Tips for Educators
Get Connected, Grow Your Professional Learning Network, and Reinvigorate Your Career
By Brad Currie, Billy Krakower, and Scott Rocco
(@bradmcurrie, @wkrakower, @ScottRRocco)

Whatever questions you have about education or about how you can be even better at your job, you'll find ideas, resources, and a vibrant network of professionals ready to help you on Twitter. In *140 Twitter Tips for Educators*, #Satchat hosts and founders of Evolving Educators, Brad Currie, Billy Krakower, and Scott Rocco offer step-by-step instructions to help you master the basics of Twitter, build an online following, and become a Twitter rock star.

Pure Genius
Building a Culture of Innovation and Taking 20% Time to the Next Level
By Don Wettrick (@DonWettrick)

For far too long, schools have been bastions of boredom, killers of creativity, and way too comfortable with compliance and conformity. In *Pure Genius*, Don Wettrick explains how collaboration—with experts, students, and other educators—can help you create interesting, and even life-changing, opportunities for learning. Wettrick's book inspires and equips educators with a systematic blueprint for teaching innovation in any school.

The Innovator's Mindset
Empower Learning, Unleash Talent, and Lead a Culture of Creativity
By George Couros (@gcouros)

The traditional system of education requires students to hold their questions and compliantly stick to the scheduled curriculum. But our job as educators is to provide new and better opportunities for our students. It's time to recognize that compliance doesn't foster innovation, encourage critical thinking, or inspire creativity—and those are the skills our students need to succeed. In *The Innovator's Mindset*, George Couros encourages teachers and administrators to empower their learners to wonder, to explore—and to become forward-thinking leaders.

Ditch That Textbook
Free Your Teaching and Revolutionize Your Classroom
By Matt Miller (@jmattmiller)

Textbooks are symbols of centuries of old education. They're often outdated as soon as they hit students' desks. Acting "by the textbook" implies compliance and a lack of creativity. It's time to ditch those textbooks—and those textbook assumptions about learning! In *Ditch That Textbook*, teacher and blogger Matt Miller encourages educators to throw out meaningless, pedestrian teaching and learning practices. He empowers them to evolve and improve on old, standard, teaching methods. *Ditch That Textbook* is a support system, toolbox, and manifesto to help educators free their teaching and revolutionize their classrooms.

Master the Media
How Teaching Media Literacy Can Save Our Plugged-in World
By Julie Smith (@julnilsmith)

Written to help teachers and parents educate the next generation, *Master the Media* explains the history, purpose, and messages behind the media. The point isn't to get kids to unplug; it's to help them make informed choices, understand the difference between truth and lies, and discern perception from reality. Critical thinking leads to smarter decisions—and it's why media literacy can save the world.

Your School Rocks ... So Tell People!
Passionately Pitch and Promote the Positives Happening on Your Campus
By Ryan McLane and Eric Lowe (@McLane_Ryan, @EricLowe21)

Great things are happening in your school every day. The problem is: no one beyond your school walls knows about them. School principals Ryan McLane and Eric Lowe want to help you get the word out! *In Your School Rocks...So Tell People!*, McLane and Lowe offer more than seventy immediately actionable tips along with easy-to-follow instructions and links to video tutorials. This practical guide will equip you to create an effective and manageable communication strategy using social media tools. Learn how to keep your students' families and community connected, informed, and excited about what's going on in your school.

THE ZEN TEACHER
Creating FOCUS, SIMPLICITY, and TRANQUILITY in the Classroom
By Dan Tricarico (@TheZenTeacher)

Teachers have incredible power to influence, even improve, the future. In *The Zen Teacher*, educator, blogger, and speaker Dan Tricarico provides practical, easy-to-use techniques to help teachers be their best—unrushed and fully focused—so they can maximize their performance and improve their quality of life. In this introductory guide, Dan Tricarico explains what it means to develop a Zen practice—something that has nothing to do with religion and everything to do with your ability to thrive in the classroom.

PLAY LIKE A PIRATE
Engage Students with Toys, Games, and Comics
By Quinn Rollins (@jedikermit)

Yes! School can be simultaneously fun and educational. *In Play Like a Pirate*, Quinn Rollins offers practical, engaging strategies and resources that make it easy to integrate fun into your curriculum. Regardless of the grade level you teach, you'll find inspiration and ideas that will help you engage your students in unforgettable ways.

EXPLORE LIKE A PIRATE
Gamification and Game-Inspired Course Design to Engage, Enrich, and Elevate Your Learners
By Michael Matera (@MrMatera)

Are you ready to transform your classroom into an experiential world that flourishes on collaboration and creativity? Then set sail with classroom game designer and educator, Michael Matera, as he reveals the possibilities and power of game-based learning. In *eXPlore Like a Pirate*, Matera serves as your experienced guide to help you apply the most motivational techniques of gameplay to your classroom. You'll learn gamification strategies that will work with and enhance (rather than replace) your current curriculum and discover how these engaging methods can be applied to any grade level or subject.

LEARN LIKE A *PIRATE*
Empower Your Students to Collaborate, Lead, and Succeed
By Paul Solarz (@PaulSolarz)

Today's job market demands that students be prepared to take responsibility for their lives and careers. We do them a disservice if we teach them how to earn passing grades without equipping them to take charge of their education. In *Learn Like a Pirate*, Paul Solarz explains how to design classroom experiences that encourage students to take risks and explore their passions in a stimulating, motivating, and supportive environment where improvement, rather than grades, is the focus. Discover how student-led classrooms help students thrive and develop into self-directed, confident citizens who are capable of making smart, responsible decisions, all on their own.

THE CLASSROOM CHEF
Sharpen your lessons. Season your classes. Make math meaningful.
By John Stevens and Matt Vaudrey
(@Jstevens009, @MrVaudrey)

In *The Classroom Chef*, math teachers and instructional coaches John Stevens and Matt Vaudrey share their secret recipes, ingredients, and tips for serving up lessons that engage students and help them "get" math. You can use these ideas and methods as-is, or better yet, tweak them and create your own enticing educational meals. The message the authors share is that, with imagination and preparation, *every* teacher can be a Classroom Chef.

HOW MUCH WATER DO WE HAVE?
5 Success Principles for Conquering Any Change and Thriving in Times of Change
By Pete Nunweiler with Kris Nunweiler

In *How Much Water Do We Have?* Pete Nunweiler identifies five key elements—information, planning, motivation, support, and leadership—that are necessary for the success of any goal, life transition, or challenge. Referring to these elements as the 5 Waters of Success, Pete explains that like the water we drink, you need them to thrive in today's rapidly paced world. If you're feeling stressed out, overwhelmed, or uncertain at work or at home, pause and look for the signs of dehydration. Learn how to find, acquire, and use the 5 Waters of Success—so you can share them with your team and family members.

About the Authors

Photo by Alex Kang

Alice Keeler has a Masters degree in Educational Media Design and Technology, is a Google Certified Teacher, New Media Consortium K12 Ambassador, Microsoft Innovative Educator, and is LEC Admin, Online, and Blended certified. She taught high school math for fourteen years and currently serves as Adjunct Professor of Curriculum Instruction and Technology at California State University, Fresno. Alice Keeler has developed and taught online K12 courses, as well as the Innovative Educator Advanced Studies Certificate. She has worked on the YouTube teachers project and the Google Play for Education project. She served on the New Media Consortium Horizon Report advisory panel for 2013, 2014, and 2015. She is a Bing in the Classroom lesson developer. A believer in the importance of connectivity, she founded #coffeeEDU (coffeeEDU.org) and #profchat.

Twitter: @alicekeeler

Libbi R. Miller, Ed.D. serves as Assistant Professor of Curriculum, Instruction and Technology at California State University, Fresno. In her courses, she focuses on preparing future teachers to integrate technology and create engaging student-centered classrooms. Prior to moving into higher education, Libbi taught middle school and alternative education. Libbi's  research involves investigating educational technology tools for its intersection with democratic and socially just education. She believes that, when used correctly, technology can become a tool for empowerment and student voice. Technology can move education outside of the classroom walls and connect students with learners, professionals, and activists around the world.

Twitter: @MillerLibbi

CPSIA information can be obtained
at www.ICGtesting.com
Printed in the USA
BVOW07s1851160516
448297BV00023B/247/P